Ocean Animals

by Margie Burton, Cathy French, and Tammy Jones

We like going to the ocean.

We see many animals when we go to the ocean. Some of the animals live in the ocean and some of the animals live beside the ocean.

Some ocean animals live deep in the water. The water is so deep that there is no light. They do not come up to the top of the ocean.

We have to use a light to see the octopus deep in the water.

Some animals live in the water at the top of the ocean.
They jump out of the water.
They like to play.

There are many kinds of fish in the ocean.

Some of the animals that live in the ocean have fins to help them swim.

Their tails help them swim, too.

Some ocean animals have flippers to help them swim in the ocean.

flipper

tail

Some ocean animals use their tails like flippers.

Some animals live by the ocean.
This bird lives by the ocean.
He looks for food to catch.

The pelican keeps his food in his mouth until he wants to eat.

This bird lives by the ocean, too.
He looks for food to catch
by the ocean.

The sandpiper uses his beak to look for food
in the sand and in the water.

Some of the ocean animals live in the sand by the ocean. We like to dig them out of the sand.

We find shells in the sand, too.

Some shells are homes for animals who live in the ocean and beside the ocean.

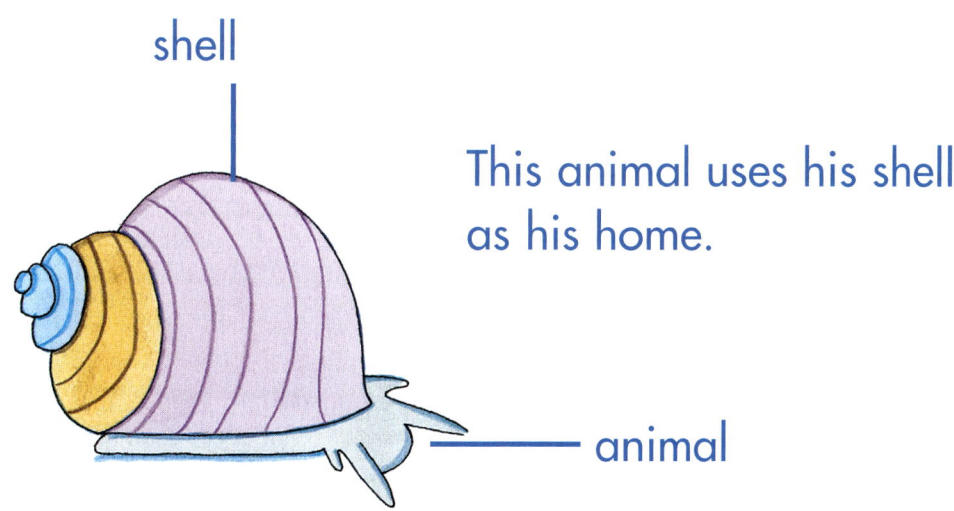

The ocean has many kinds of animals.

These are some of the animals that live in the ocean.